I0076996

Florian & Relly
Ion Victoria
PETRESCU PETRESCU

THE DESIGN OF GEARINGS WITH HIGH EFFICIENCY

Lulu. Lulu. Lulu. Lulu. Lulu.

Publisher

London Uk   **2011**   London UK

Scientific reviewers:

Prof. Consul. Dr. Ing. Păun ANTONESCU

Prof. Dr. Ing. Adriana COMĂNESCU

Copyright

Title book: THE DESIGN OF GEARINGS WITH HIGH EFFICIENCY

Authors book: Florian Ion PETRESCU & Relly Victoria PETRESCU

© 2011, Florian PETRESCU & Relly PETRESCU

petrescuflorian@yahoo.com

ALL RIGHTS RESERVED. This book contains material protected under International and Federal Copyright Laws and Treaties. Any unauthorized reprint or use of this material is prohibited. No part of this book may be reproduced or transmitted in any form or by any means, electronic or mechanical, including photocopying, recording, or by any information storage and retrieval system without express written permission from the authors / publisher. Lulu.

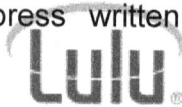

ISBN 978-1-4467-9054-0

A Short Book Description:

Development and diversification of machines and mechanisms with applications in all areas of scientific research requires new systematization and improvement of existing mechanical systems by creating new mechanisms adapted to the modern requirements, which involve more complex topological structures. Modern industry, the practice of engineering design and manufacture increasingly rely more on scientific research results and practical.

The processes of robotisation of today define and influence the emergence of new industries, with applications in specific environmental conditions, handling of objects in outer space, and are leading teleoperator in disciplines such as medicine, automations, nuclear energetic, etc.

In this context this paper attempts to bring a contribution to science and technology applied in the kinematic and dynamic analysis and synthesis of mechanisms with gearings.

The book presents an original method to determine the efficiency of the gear. The originality of this method relies on the eliminated friction modulus. The work is analyzing the influence of a few parameters concerning gear efficiency. These parameters are: z_1 - the number of teeth for the primary wheel of gear; z_2 - the number of teeth of the secondary wheel of gear; α_0 - the normal pressure angle on the divided circle; β - the inclination angle. With the relations presented in this paper, one can synthesize the gear's mechanisms.

We begin with the right teeth (the toothed gear), with i=-4, once for z_1 we shall take successively different values, rising from 8 teeth. One can see that for 8 teeth of the driving wheel the standard pressure angle, $\alpha_0=20^0$, is too small to be used (it obtains a minimum pressure angle, α_m, negative and this fact is not admitted; see the first table). In the second table we shall diminish (in module) the value for the ratio of transmission, i, from 4 to 2. We will see how for a lower value of the number of teeth of the wheel 1, the standard pressure angle ($\alpha_0=20^0$) is too small and it will be necessary to increase it to a minimum value. For example, if $z_1=8$, the necessary minimum value is $\alpha_0=290$ for an i=-4 (see the table 1) and $\alpha_0=28^0$ for an i=-2 (see the table 2). If $z_1=10$, the necessary minimum pressure angle is $\alpha_0=26^0$ for i=-4 (see the table 1) and $\alpha_0=25^0$ for i=-2 (see the table 2). When the number of teeth of the wheel 1 increases, we can decrease the normal pressure angle, α_0. We will see that for $z_1=90$ it can take a less value for the normal pressure angle (for the pressure angle of reference), $\alpha_0=8^0$. In the table 3 we increases the module of i value (the ratio of transmission), from 2 to 6.

In the table 4, the teeth are bended ($\beta \neq 0$). The module i takes now the value 2.

The efficiency (of the gear) increases when the number of teeth for the driving wheel 1, z_1, increases, and when the pressure angle, α_0, diminishes; z_2 and i_{12} have not so much influence about the efficiency value.

We can easily see that for the value $\alpha_0 = 20^0$, the efficiency takes roughly the value $\eta \approx 0.89$ for any values of the others parameters (this justifies the choice of this value, $\alpha_0 = 20^0$, for the standard pressure angle of reference).

But the better efficiency may be obtained only for a $\alpha_0 \neq 20^0$ ($\alpha_0 < 20^0$).

The pressure angle of reference, α_0, can be decreased, when in the same time, the number of teeth for the driving wheel 1, z_1, increases, to increase the gears' efficiency.

Contrary, when we desire to create a gear with a low z_1 (for a less gauge), it will be necessary to increase the α_0 value, for maintaining a positive value for α_m (in this case the gear efficiency will be diminished).

When β increases, the efficiency (η) increases too, but its growth is insignificant. We can see in the last part of the work, that in reality it (β increases) produces a decrease in yield.

The module of the gear, m, has not any influence on the gear's efficiency value.

When α_0 is diminished one can take a higher normal module, for increasing the addendum of teeth, but the increase of the m at the same time with the increase of the z_1 can lead to a greater gauge.

The gears' efficiency (η) is really a function of α_0 and z_1: $\eta=f(\alpha_0,z_1)$; the two angles (α_m and α_M) are just the intermediate parameters (intermediate variables).

For a good projection of the gear, it's necessary a z_1 and a z_2 greater than 30-60; but this condition may increase the gauge of mechanism; when the numbers of teeth z_1 and z_2 beyond the 30 value, the efficiency of the gearing are greater, and the values of the two different efficiencies leveled; this can be a great advantage in transmissions, especially in planetary transmissions, where the moments may come from both directions; will result a better and more equilibrated functionality (But these are the subject of a future work).

In the second (and last) part the book presents shortly an original method to obtain the efficiency of the geared transmissions in function of the contact ratio. With the presented relations one can make the dynamic synthesis of the geared transmissions having in view increasing the efficiency of gearing mechanisms in work (the accuracy of calculations will be high).

One calculates the efficiency of a geared transmission, having in view the fact that at one moment there are several couples of teeth in contact, and not just one.

The start model has got four pairs of teeth in contact (4 couples) concomitantly.

The first couple of teeth in contact has the contact point i, defined by the ray r_{i1}, and the pressure angle α_{i1}; the forces which act at this point are: the motor force F_{mi}, perpendicular to the position vector r_{i1} at i and the force transmitted from the wheel 1 to the wheel 2 through the point i, $F_{\tau i}$, parallel to the path of action and with the sense from the wheel 1 to the wheel 2, the transmitted force being practically the projection of the motor force on the path of action; the defined velocities are similar to the forces (having in view the original kinematics, or the precise kinematics adopted); the same parameters will be defined for the next three points of contact, j, k, l (see fig. 2).

The best efficiency can be obtained with the internal gearing when the drive wheel 1 is the ring; the minimum efficiency will be obtained when the drive wheel 1 of the internal gearing has external teeth. For the external gearing, the best efficiency is obtained when the bigger wheel is the drive wheel; *when one decreases the normal angle* α_0, *the contact ratio increases and the efficiency increases as well.* The efficiency increases too, when the number of teeth of the drive wheel 1 increases (when z_1 increases).

Generally we use gearings with teeth inclined (with bended teeth). For gears with bended teeth, the calculations show a decrease in yield when the inclination angle increases. For angles with inclination which not exceed 25 degree the efficiency of gearing is good (see the table 6). When the inclination angle (β) exceeds 25 degrees the gearing will suffer a significant drop in yield (see the tables 7 and 8).

The calculation relationships (33-35) are general (Have a general nature). They have the advantage that can be used with great precision in determining the efficiency of any type of gearings.

1 Introduction

In this paper the authors present an original method to calculating the efficiency of the gear.

The originality consists in the way of determination of the gear's efficiency because one hasn't used the friction forces of couple (this new way eliminates the classical method). One eliminates the necessity of determining the friction coefficients by different experimental methods as well. The efficiency determined by the new method is the same like the classical efficiency, namely the mechanical efficiency of the gear.

Some mechanisms work by pulses and are transmitting the movement from an element to another by pulses and not by friction. Gears work practically only by pulses. The component of slip or friction is practically the loss. Because of this the mechanical efficacy becomes practically the mechanical efficiency of gear.

The paper is analyzing the influence of a few parameters concerning gear efficiency. With the relations presented in this paper, one can synthesize the gear's mechanisms. Today, the gears are present every where in the mechanical's world.

2 Determining the Momentary Mechanical Efficiency

The calculating relations are the next (1-20), (see the fig. 1):

$$\begin{cases} F_\tau = F_m \cdot \cos\alpha_1 & F_\psi = F_m \cdot \sin\alpha_1 & \overline{F}_m = \overline{F}_\tau + \overline{F}_\psi \\ v_2 = v_1 \cdot \cos\alpha_1 & v_{12} = v_1 \cdot \sin\alpha_1 & \overline{v}_1 = \overline{v}_2 + \overline{v}_{12} \end{cases} \qquad (1)$$

With: F_m - the motive force (the driving force); F_τ - the transmitted force (the useful force); F_ψ - the slide force (the lost force); v_1 - the velocity of element 1, or the speed of wheel 1 (the driving wheel); v_2 - the velocity of element 2, or the speed of wheel 2 (the driven wheel); v_{12} - the relative speed of the wheel 1 in relation with the wheel 2 (this is a sliding speed).

The consumed power (in this case the driving power) takes the form (2). The useful power (the transmitted power from the profile 1 to the profile 2) will be written in the relation (3). The lost power will be written in the form (4).

$$P_c \equiv P_m = F_m \cdot v_1 \qquad (2)$$

$$P_u \equiv P_\tau = F_\tau \cdot v_2 = F_m \cdot v_1 \cdot \cos^2 \alpha_1 \qquad (3)$$

$$P_\psi = F_\psi \cdot v_{12} = F_m \cdot v_1 \cdot \sin^2 \alpha_1 \qquad (4)$$

The momentary efficiency of couple will be calculated directly with the next relation:

$$\left\{ \eta_i = \frac{P_u}{P_c} \equiv \frac{P_\tau}{P_m} = \frac{F_m \cdot v_1 \cdot \cos^2 \alpha_1}{F_m \cdot v_1} \right. \quad \eta_i = \cos^2 \alpha_1 \qquad (5)$$

The momentary losing coefficient will be written in the form (6):

$$\left\{ \begin{array}{l} \psi_i = \dfrac{P_\psi}{P_m} = \dfrac{F_m \cdot v_1 \cdot \sin^2 \alpha_1}{F_m \cdot v_1} = \sin^2 \alpha_1 \\ \eta_i + \psi_i = \cos^2 \alpha_1 + \sin^2 \alpha_1 = 1 \end{array} \right. \qquad (6)$$

One can easily see that the sum of the momentary efficiency and the momentary losing coefficient is 1.

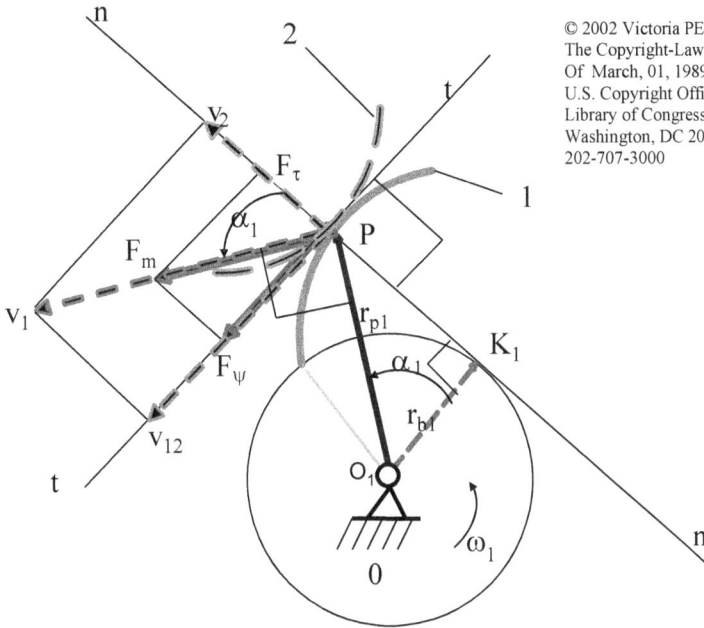

© 2002 Victoria PETRESCU
The Copyright-Law
Of March, 01, 1989
U.S. Copyright Office
Library of Congress
Washington, DC 20559-6000
202-707-3000

Fig. 1 *The forces of the gear*

Now one can determine the geometrical elements of the gear. These elements will be used in determining of the couple efficiency, η.

3 The Geometrical Elements of the Gear

One can determine the next geometrical elements of the external gear (for the right teeth, $\beta=0$): The radius of the basic circle of wheel 1 (of the driving wheel) (7); the radius of the outside circle of wheel 1 (8); the maximum pressure angle of the gear (9):

$$r_{b1} = \frac{1}{2} \cdot m \cdot z_1 \cdot \cos \alpha_0 \qquad (7)$$

$$r_{a1} = \frac{1}{2} \cdot (m \cdot z_1 + 2 \cdot m) = \frac{m}{2} \cdot (z_1 + 2) \qquad (8)$$

$$\cos \alpha_{1M} = \frac{r_{b1}}{r_{a1}} = \frac{\frac{1}{2} \cdot m \cdot z_1 \cdot \cos \alpha_0}{\frac{1}{2} \cdot m \cdot (z_1 + 2)} = \frac{z_1 \cdot \cos \alpha_0}{z_1 + 2} \qquad (9)$$

And now one determines the same parameters for the wheel 2: the radius of basic circle (10) and the radius of the outside circle (11).

$$r_{b2} = \frac{1}{2} \cdot m \cdot z_2 \cdot \cos \alpha_0 \qquad (10)$$

$$r_{a2} = \frac{m}{2} \cdot (z_2 + 2) \qquad (11)$$

Now one can determine the minimum pressure angle of the external gear (12):

$$tg\,\alpha_{1m} = [(z_1 + z_2) \cdot \sin \alpha_0 - \sqrt{z_2^2 \cdot \sin^2 \alpha_0 + 4 \cdot z_2 + 4}]/(z_1 \cdot \cos \alpha_0) \qquad (12)$$

Now one can determine, for the external gear, the minimum (12) and the maximum (9) pressure angle for the right teeth. For the external gear with bended teeth ($\beta \neq 0$) one uses the relations (13, 14 and 15):

$$tg\,\alpha_t = \frac{tg\,\alpha_0}{\cos \beta} \qquad (13)$$

$$tg\,\alpha_{1m} = \left[(z_1 + z_2)\cdot\frac{\sin\alpha_t}{\cos\beta} - \sqrt{z_2^2\cdot\frac{\sin^2\alpha_t}{\cos^2\beta} + 4\cdot\frac{z_2}{\cos\beta} + 4}\,\right]\cdot\frac{\cos\beta}{z_1\cdot\cos\alpha_t} \qquad (14)$$

$$\cos\alpha_{1M} = \frac{\dfrac{z_1\cdot\cos\alpha_t}{\cos\beta}}{\dfrac{z_1}{\cos\beta} + 2} \qquad (15)$$

For the internal gear with bended teeth (β≠0) one uses the relations (13 with 16, 17-A or with 18, 19-B).

A. When the driving wheel 1, has external teeth:

$$tg\,\alpha_{1m} = \frac{[(z_1 - z_2)\cdot\dfrac{\sin\alpha_t}{\cos\beta} + \sqrt{z_2^2\cdot\dfrac{\sin^2\alpha_t}{\cos^2\beta} - 4\cdot\dfrac{z_2}{\cos\beta} + 4}]\cdot\dfrac{\cos\beta}{z_1\cdot\cos\alpha_t}}{} \qquad (16)$$

$$\cos\alpha_{1M} = \frac{\dfrac{z_1\cdot\cos\alpha_t}{\cos\beta}}{\dfrac{z_1}{\cos\beta} + 2} \qquad (17)$$

B. When the driving wheel 1, have internal teeth:

$$tg\,\alpha_{1M} = \frac{[(z_1 - z_2)\cdot\dfrac{\sin\alpha_t}{\cos\beta} + \sqrt{z_2^2\cdot\dfrac{\sin^2\alpha_t}{\cos^2\beta} + 4\cdot\dfrac{z_2}{\cos\beta} + 4}]\cdot\dfrac{\cos\beta}{z_1\cdot\cos\alpha_t}}{} \qquad (18)$$

$$\cos \alpha_{1m} = \frac{\dfrac{z_1 \cdot \cos \alpha_t}{\cos \beta}}{\dfrac{z_1}{\cos \beta} - 2} \qquad (19)$$

4 Determination of the Efficiency

The efficiency of the gear will be calculated through the integration of momentary efficiency on all sections of gearing movement, namely from the minimum pressure angle to the maximum pressure angle; the relation (20).

$$\eta = \frac{1}{\Delta\alpha} \cdot \int_{\alpha_m}^{\alpha_M} \eta_i \cdot d\alpha = \frac{1}{\Delta\alpha} \int_{\alpha_m}^{\alpha_M} \cos^2\alpha \cdot d\alpha =$$

$$= \frac{1}{2\cdot\Delta\alpha} \cdot [\frac{1}{2}\cdot\sin(2\cdot\alpha) + \alpha]_{\alpha_m}^{\alpha_M} = \qquad (20)$$

$$= \frac{1}{2\cdot\Delta\alpha}[\frac{\sin(2\alpha_M) - \sin(2\alpha_m)}{2} + \Delta\alpha] =$$

$$= \frac{\sin(2\cdot\alpha_M) - \sin(2\cdot\alpha_m)}{4\cdot(\alpha_M - \alpha_m)} + 0.5$$

5 The Calculated Efficiency of the Gearing

We shall now see four tables with the calculated efficiency depending on the input parameters and once we proceed with the results we will draw some conclusions.

The input parameters are:

z_1 = the number of teeth for the driving wheel 1;

z_2 = the number of teeth for the driven wheel 2, or the ratio of transmission, i (i_{12}=-z_2/z_1);

a_0 = the pressure angle normal on the divided circle;

β = the bend angle.

$i_{12effective} = -4$	right teeth	Table 1
$z_1 = 8$	$z_2 = 32$	
$\alpha_0 = 20^0$?	$\alpha_0 = 29^0$	$\alpha_0 = 35^0$
$\alpha_m = -16.22^0$?	$\alpha_m = 0.7159^0$	$\alpha_m = 11.1303^0$
$\alpha_M = 41.2574^0$	$\alpha_M = 45.5974^0$	$\alpha_M = 49.0560^0$
	$\eta = 0.8111$	$\eta = 0.7308$
$z_1 = 10$	$z_2 = 40$	
$\alpha_0 = 20^0$?	$\alpha_0 = 26^0$	$\alpha_0 = 30^0$
$\alpha_m = -9.89^0$?	$\alpha_m = 1.3077^0$	$\alpha_m = 8.2217^0$
$\alpha_M = 38.4568^0$	$\alpha_M = 41.4966^0$	$\alpha_M = 43.8060^0$
	$\eta = 0.8375$	$\eta = 0.7882$
$z_1 = 18$	$z_2 = 72$	
$\alpha_0 = 19^0$	$\alpha_0 = 20^0$	$\alpha_0 = 30^0$
$\alpha_m = 0.9860^0$	$\alpha_m = 2.7358^0$	$\alpha_m = 18.2830^0$
$\alpha_M = 31.6830^0$	$\alpha_M = 32.2505^0$	$\alpha_M = 38.7922^0$
$\eta = 0.90105$	$\eta = 0.8918$	$\eta = 0.7660$
$z_1 = 30$	$z_2 = 120$	
$\alpha_0 = 15^0$	$\alpha_0 = 20^0$	$\alpha_0 = 30^0$
$\alpha_m = 1.5066^0$	$\alpha_m = 9.5367^0$	$\alpha_m = 23.1225^0$
$\alpha_M = 25.1018^0$	$\alpha_M = 28.2414^0$	$\alpha_M = 35.7181^0$
$\eta = 0.9345$	$\eta = 0.8882$	$\eta = 0.7566$
$z_1 = 90$	$z_2 = 360$	
$\alpha_0 = 8^0$?	$\alpha_0 = 9^0$	$\alpha_0 = 20^0$
$\alpha_m = -0.1638^0$?	$\alpha_m = 1.5838^0$	$\alpha_m = 16.4999^0$
$\alpha_M = 14.3637^0$	$\alpha_M = 14.9354^0$	$\alpha_M = 23.1812^0$
	$\eta = 0.9750$	$\eta = 0.8839$

$i_{12effective} = -2$	right teeth	Table 2
$z_1 = 8$	$z_2 = 16$	
$\alpha_0 = 20^0$?	$\alpha_0 = 28^0$	$\alpha_0 = 35^0$
$\alpha_m = -12.65^0$?	$\alpha_m = 0.9149^0$	$\alpha_m = 12.2933^0$
$\alpha_M = 41.2574^0$	$\alpha_M = 45.0606^0$	$\alpha_M = 49.0559^0$
	$\eta = 0.8141$	$\eta = 0.7236$
$z_1 = 10$	$z_2 = 20$	
$\alpha_0 = 20^0$?	$\alpha_0 = 25^0$	$\alpha_0 = 30^0$
$\alpha_m = -7.13^0$?	$\alpha_m = 1.3330^0$	$\alpha_m = 9.4106^0$
$\alpha_M = 38.4568^0$	$\alpha_M = 40.9522^0$	$\alpha_M = 43.8060^0$
	$\eta = 0.8411$	$\eta = 0.7817$
$z_1 = 18$	$z_2 = 36$	
$\alpha_0 = 18^0$	$\alpha_0 = 20^0$	$\alpha_0 = 30^0$
$\alpha_m = 0.6756^0$	$\alpha_m = 3.9233^0$	$\alpha_m = 18.6935^0$
$\alpha_M = 31.1351^0$	$\alpha_M = 32.2505^0$	$\alpha_M = 38.7922^0$
$\eta = 0.9052$	$\eta = 0.8874$	$\eta = 0.7633$
$z_1 = 30$	$z_2 = 60$	
$\alpha_0 = 14^0$	$\alpha_0 = 20^0$	$\alpha_0 = 30^0$
$\alpha_m = 0.8845^0$	$\alpha_m = 10.0416^0$	$\alpha_m = 23.2774^0$
$\alpha_M = 24.5427^0$	$\alpha_M = 28.2414^0$	$\alpha_M = 35.7181^0$
$\eta = 0.9388$	$\eta = 0.8859$	$\eta = 0.7555$
$z_1 = 90$	$z_2 = 180$	
$\alpha_0 = 8^0$	$\alpha_0 = 20^0$	$\alpha_0 = 30^0$
$\alpha_m = 0.5227^0$	$\alpha_m = 16.5667^0$	$\alpha_m = 27.7825^0$
$\alpha_M = 14.3637^0$	$\alpha_M = 23.1812^0$	$\alpha_M = 32.0917^0$
$\eta = 0.9785$	$\eta = 0.8836$	$\eta = 0.7507$

$i_{12effective} = -6$	right teeth	Table 3
$z_1 = 8$	$z_2 = 48$	
$\alpha_0 = 20^0$?	$\alpha_0 = 30^0$	$\alpha_0 = 35^0$
$\alpha_m = -17.86^0$?	$\alpha_m = 1.7784^0$	$\alpha_m = 10.660^0$
$\alpha_M = 41.2574^0$	$\alpha_M = 46.1462^0$	$\alpha_M = 49.0559^0$
	$\eta = 0.8026$	$\eta = 0.7337$
$z_1 = 10$	$z_2 = 60$	
$\alpha_0 = 20^0$?	$\alpha_0 = 26^0$	$\alpha_0 = 30^0$
$\alpha_m = -11.12^0$?	$\alpha_m = 0.6054^0$	$\alpha_m = 7.7391^0$
$\alpha_M = 38.4568^0$	$\alpha_M = 41.4966^0$	$\alpha_M = 43.8060^0$
	$\eta = 0.8403$	$\eta = 0.7908$
$z_1 = 18$	$z_2 = 108$	
$\alpha_0 = 19^0$	$\alpha_0 = 20^0$	$\alpha_0 = 30^0$
$\alpha_m = 0.4294^0$	$\alpha_m = 2.2449^0$	$\alpha_m = 18.1280^0$
$\alpha_M = 31.6830^0$	$\alpha_M = 32.2505^0$	$\alpha_M = 38.7922^0$
$\eta = 0.9028$	$\eta = 0.8935$	$\eta = 0.7670$
$z_1 = 30$	$z_2 = 180$	
$\alpha_0 = 15^0$	$\alpha_0 = 20^0$	$\alpha_0 = 30^0$
$\alpha_m = 1.0922^0$	$\alpha_m = 9.3414^0$	$\alpha_m = 23.0666^0$
$\alpha_M = 25.1018^0$	$\alpha_M = 28.2414^0$	$\alpha_M = 35.7181^0$
$\eta = 0.9356$	$\eta = 0.8891$	$\eta = 0.7570$
$z_1 = 90$	$z_2 = 540$	
$\alpha_0 = 9^0$	$\alpha_0 = 20^0$	$\alpha_0 = 30^0$
$\alpha_m = 1.3645^0$	$\alpha_m = 16.4763^0$	$\alpha_m = 27.7583^0$
$\alpha_M = 14.9354^0$	$\alpha_M = 23.1812^0$	$\alpha_M = 32.0917^0$
$\eta = 0.9754$	$\eta = 0.8841$	$\eta = 0.7509$

We begin with the right teeth (the toothed gear), with i=-4, once for z_1 we shall take successively different values, rising from 8 teeth.

One can see that for 8 teeth of the driving wheel the standard pressure angle, $a_0=20^0$, is to small to be used (one obtains a minimum pressure angle, a_m, negative and this fact is not admitted!).

In the second table we shall diminish (in module) the value for the ratio of transmission, i, from 4 to 2.

We will see now, how for a lower value of the number of teeth of the wheel 1, the standard pressure angle ($\alpha_0=20^0$) is to small and it will be necessary to increase it to a minimum value.

For example, if $z_1=8$, the necessary minimum value is $\alpha_0=29^0$ for an i=-4 (see the table 1) and $\alpha_0=28^0$ for an i=-2 (see the table 2).

If $z_1=10$, the necessary minimum pressure angle is $\alpha_0=26^0$ for i=-4 (see the table 1) and $\alpha_0=25^0$ for i=-2 (see the table 2).

When the number of teeth of the wheel 1 increases, one can decrease the normal pressure angle, α_0. One shall see that for $z_1=90$ one can take less for the normal pressure angle (for the pressure angle of reference), $\alpha_0=8^0$. In the table 3 one increases the module of i, value (for the ratio of transmission), from 2 to 6.

In the table 4, the teeth are bended ($\beta\neq0$).

$i_{12effective}= - 4$	bend teeth	Table 4
	$\beta=15^0$	
$z_1=8$	$z_2=32$	
$\alpha_0=20^0$?	$\alpha_0=30^0$	$\alpha_0=35^0$
$\alpha_m=-16.836^0$?	$\alpha_m=1.1265^0$	$\alpha_m=9.4455^0$
$\alpha_M=41.0834^0$	$\alpha_M=46.2592^0$	$\alpha_M=49.2953^0$
	$\eta=0.8046$	$\eta=0.7390$
$z_1=10$	$z_2=40$	
$\alpha_0=20^0$?	$\alpha_0=26^0$	$\alpha_0=30^0$
$\alpha_m=-10.563^0$?	$\alpha_m=0.2355^0$	$\alpha_m=6.9188^0$
$\alpha_M=38.3474^0$	$\alpha_M=41.57139^0$	$\alpha_M=43.9965^0$
	$\eta=0.8412$	$\eta=0.7937$
$z_1=18$	$z_2=72$	
$\alpha_0=19^0$	$\alpha_0=20^0$	$\alpha_0=30^0$
$\alpha_m=0.32715^0$	$\alpha_m=2.0283^0$	$\alpha_m=17.1840^0$
$\alpha_M=31.7180^0$	$\alpha_M=32.3202^0$	$\alpha_M=39.1803^0$
$\eta=0.9029$	$\eta=0.8938$	$\eta=0.7702$
$z_1=30$	$z_2=120$	
$\alpha_0=15^0$	$\alpha_0=20^0$	$\alpha_0=30^0$
$\alpha_m=1.0269^0$	$\alpha_m=8.8602^0$	$\alpha_m=22.1550^0$
$\alpha_M=25.1344^0$	$\alpha_M=28.4591^0$	$\alpha_M=36.2518^0$
$\eta=0.9357$	$\eta=0.8899$	$\eta=0.7593$
$z_1=90$	$z_2=360$	
$\alpha_0=9^0$	$\alpha_0=20^0$	$\alpha_0=30^0$
$\alpha_m=1.3187^0$	$\alpha_m=15.8944^0$	$\alpha_m=26.9403^0$
$\alpha_M=14.9648^0$	$\alpha_M=23.6366^0$	$\alpha_M=32.8262^0$
$\eta=0.9754$	$\eta=0.8845$	$\eta=0.7513$

6. Observations

The efficiency (of the gear) increases when the number of teeth for the driving wheel 1, z_1, increases too and when the pressure angle, a_0, diminishes; z_2 or i_{12} are not so much influence about the efficiency value.

One can easily see that for the value $a_0=20^0$, the efficiency takes roughly the value $\eta \approx 0.89$ for any values of the others parameters (this justifies the choice of this value, $a_0 = 20^0$, for the standard pressure angle of reference).

The better efficiency may be obtained only for a $a_0 \neq 20^0$. But the pressure angle of reference, a_0, can be decreased the same time the number of teeth for the driving wheel 1, z_1, increases, to increase the gears' efficiency.

Contrary, when we desire to create a gear with a low z_1 (for a less gauge), it will be necessary to increase the a_0 value, for maintaining a positive value for a_m (in this case the gear efficiency will be diminished).

When β increases, the efficiency, η, increases too, but the growth is insignificant. The module of the gear, m, has not any influence on the gear's efficiency value.

When a_0 is diminished one can take a higher normal module, for increasing the addendum of teeth, but the increase of the m at the same time with the increase of the z_1 can lead to a greater gauge.

The gears' efficiency, η, is really a function of a_0 and z_1: $\eta=f(a_0,z_1)$; a_m and a_M are just the intermmediate parameters.

For a good projection of the gear, it's necessary a z_1 and a z_2 greater than 30-60; but this condition may increase the gauge of mechanism.

In this paper, one determines precisely, the dynamics-efficiency, but at the gears transmissions, the dynamics efficiency is the same like the mechanical efficiency; this is a greater advantage of the gears transmissions.

This advantage, specifically of the gear's mechanisms, may be found at the cams mechanisms with plate followers too.

7 Determining of Gearing Efficiency in Function of the Contact Ratio

One calculates the efficiency of a geared transmission, having in view the fact that at one moment there are several couples of teeth in contact, and not just one.

The start model has got four pairs of teeth in contact (4 couples) concomitantly.

The first couple of teeth in contact has the contact point i, defined by the ray r_{i1}, and the pressure angle α_{i1}; the forces which act at this point are: the motor force F_{mi}, perpendicular to the position vector r_{i1} at i and the force transmitted from the wheel 1 to the wheel 2 through the point i, $F_{\tau i}$, parallel to the path of action and with the sense from the wheel 1 to the wheel 2, the transmitted force being practically the projection of the motor force on the path of action; the defined velocities are similar to the forces (having in view the original kinematics, or the precise kinematics adopted); the same parameters will be defined for the next three points of contact, j, k, l (Fig. 2).

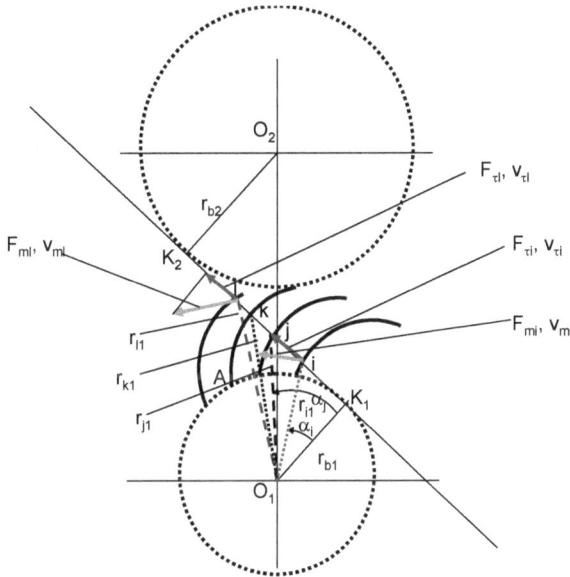

Fig. 2 *Four pairs of teeth in contact concomitantly*

For starting we write the relations between the velocities (21):

$$v_{ti} = v_{mi} \cdot \cos \alpha_i = r_i \cdot \omega_1 \cdot \cos \alpha_i = r_{b1} \cdot \omega_1$$
$$v_{tj} = v_{mj} \cdot \cos \alpha_j = r_j \cdot \omega_1 \cdot \cos \alpha_j = r_{b1} \cdot \omega_1$$
$$v_{tk} = v_{mk} \cdot \cos \alpha_k = r_k \cdot \omega_1 \cdot \cos \alpha_k = r_{b1} \cdot \omega_1 \tag{21}$$
$$v_{tl} = v_{ml} \cdot \cos \alpha_l = r_l \cdot \omega_1 \cdot \cos \alpha_l = r_{b1} \cdot \omega_1$$

From relations (21), one obtains the equality of the tangential velocities (22), and makes explicit the motor velocities (23):

$$v_{ti} = v_{tj} = v_{tk} = v_{tl} = r_{b1} \cdot \omega_1 \tag{22}$$

$$v_{mi} = \frac{r_{b1} \cdot \omega_1}{\cos \alpha_i}; v_{mj} = \frac{r_{b1} \cdot \omega_1}{\cos \alpha_j};$$

$$v_{mk} = \frac{r_{b1} \cdot \omega_1}{\cos \alpha_k}; v_{ml} = \frac{r_{b1} \cdot \omega_1}{\cos \alpha_l}$$

$$(23)$$

The forces transmitted concomitantly at the four points must be the same (24):

$$F_{ti} = F_{tj} = F_{tk} = F_{tl} = F_\tau \qquad (24)$$

The motor forces are (25):

$$F_{mi} = \frac{F_\tau}{\cos \alpha_i}; F_{mj} = \frac{F_\tau}{\cos \alpha_j};$$

$$F_{mk} = \frac{F_\tau}{\cos \alpha_k}; F_{ml} = \frac{F_\tau}{\cos \alpha_l}$$

$$(25)$$

The momentary efficiency can be written in the form (26).

$$\eta_i = \frac{P_u}{P_c} = \frac{P_\tau}{P_m} = \frac{F_{\tau i} \cdot v_{\tau i} + F_{\tau j} \cdot v_{\tau j} + F_{\tau k} \cdot v_{\tau k} + F_{\tau l} \cdot v_{\tau l}}{F_{mi} \cdot v_{mi} + F_{mj} \cdot v_{mj} + F_{mk} \cdot v_{mk} + F_{ml} \cdot v_{ml}} =$$

$$= \frac{4 \cdot F_\tau \cdot r_{b1} \cdot \omega_1}{\dfrac{F_\tau \cdot r_{b1} \cdot \omega_1}{\cos^2 \alpha_i} + \dfrac{F_\tau \cdot r_{b1} \cdot \omega_1}{\cos^2 \alpha_j} + \dfrac{F_\tau \cdot r_{b1} \cdot \omega_1}{\cos^2 \alpha_k} + \dfrac{F_\tau \cdot r_{b1} \cdot \omega_1}{\cos^2 \alpha_l}} = \qquad (26)$$

$$= \frac{4}{\dfrac{1}{\cos^2 \alpha_i} + \dfrac{1}{\cos^2 \alpha_j} + \dfrac{1}{\cos^2 \alpha_k} + \dfrac{1}{\cos^2 \alpha_l}} =$$

$$= \frac{4}{4 + tg^2 \alpha_i + tg^2 \alpha_j + tg^2 \alpha_k + tg^2 \alpha_l}$$

Relations (27) and (28) are auxiliary:

$$K_1 i = r_{b1} \cdot tg \alpha_i; K_1 j = r_{b1} \cdot tg \alpha_j; K_1 k = r_{b1} \cdot tg \alpha_k; K_1 l = r_{b1} \cdot tg \alpha_l$$

$$K_1 j - K_1 i = r_{b1} \cdot (tg \alpha_j - tg \alpha_i); K_1 j - K_1 i = r_{b1} \cdot \frac{2 \cdot \pi}{z_1} \Rightarrow tg \alpha_j = tg \alpha_i + \frac{2 \cdot \pi}{z_1} \qquad (27)$$

$$K_1 k - K_1 i = r_{b1} \cdot (tg \alpha_k - tg \alpha_i); K_1 k - K_1 i = r_{b1} \cdot 2 \cdot \frac{2 \cdot \pi}{z_1} \Rightarrow tg \alpha_k = tg \alpha_i + 2 \cdot \frac{2 \cdot \pi}{z_1}$$

$$K_1 l - K_1 i = r_{b1} \cdot (tg \alpha_l - tg \alpha_i); K_1 l - K_1 i = r_{b1} \cdot 3 \cdot \frac{2 \cdot \pi}{z_1} \Rightarrow tg \alpha_l = tg \alpha_i + 3 \cdot \frac{2 \cdot \pi}{z_1}$$

$$tg\alpha_j = tg\alpha_i \pm \frac{2\cdot\pi}{z_1};$$

$$tg\alpha_k = tg\alpha_i \pm 2\cdot\frac{2\cdot\pi}{z_1}; \qquad\qquad (28)$$

$$tg\alpha_l = tg\alpha_i \pm 3\cdot\frac{2\cdot\pi}{z_1}$$

One keeps relations (28), with the sign plus (+) for the gearing where the drive wheel 1 has external teeth (at the external or internal gearing) and with the sign (-) for the gearing where the drive wheel 1, has internal teeth (the drive wheel is a ring, only for the internal gearing). The relation of the momentary efficiency (26) uses the auxiliary relations (28) and takes the form (29).

$$\eta_i = \frac{4}{4 + tg^2\alpha_i + tg^2\alpha_j + tg^2\alpha_k + tg^2\alpha_l} =$$

$$= \frac{4}{4 + tg^2\alpha_i + (tg\alpha_i \pm \frac{2\pi}{z_1})^2 + (tg\alpha_i \pm 2 \cdot \frac{2\pi}{z_1})^2 + (tg\alpha_i \pm 3 \cdot \frac{2\pi}{z_1})^2} =$$

$$= \frac{4}{4 + 4 \cdot tg^2\alpha_i + \frac{4\pi^2}{z_1^2} \cdot (0^2 + 1^2 + 2^2 + 3^2) \pm 2 \cdot tg\alpha_i \cdot \frac{2\pi}{z_1} \cdot (0 + 1 + 2 + 3)} =$$

$$= \frac{1}{1 + tg^2\alpha_i + \frac{4\pi^2}{E \cdot z_1^2} \cdot \sum_{i=1}^{E}(i-1)^2 \pm 2 \cdot tg\alpha_i \cdot \frac{2\pi}{E \cdot z_1} \cdot \sum_{i=1}^{E}(i-1)} =$$

$$= \frac{1}{1 + tg^2\alpha_1 + \frac{4\pi^2}{E \cdot z_1^2} \cdot \frac{E \cdot (E-1) \cdot (2 \cdot E - 1)}{6} \pm \frac{4\pi \cdot tg\alpha_1}{E \cdot z_1} \cdot \frac{E \cdot (E-1)}{2}} =$$

$$= \frac{1}{1 + tg^2\alpha_1 + \frac{2\pi^2 \cdot (E-1) \cdot (2E-1)}{3 \cdot z_1^2} \pm \frac{2\pi \cdot tg\alpha_1 \cdot (E-1)}{z_1}} = \quad (29)$$

$$= \frac{1}{1 + tg^2\alpha_1 + \frac{2\pi^2}{3 \cdot z_1^2} \cdot (\varepsilon_{12} - 1) \cdot (2 \cdot \varepsilon_{12} - 1) \pm \frac{2\pi \cdot tg\alpha_1}{z_1} \cdot (\varepsilon_{12} - 1)}$$

In expression (29) one starts with relation (26) where four pairs are in contact concomitantly, but then one generalizes the expression, replacing the 4 figure (four pairs) with E couples, replacing figure 4 with the E variable, which represents the whole number of the contact ratio +1, and after restricting the sums expressions, we replace the variable E with the contact ratio ε_{12}, as well.

The mechanical efficiency offers more advantages than the momentary efficiency, and will be calculated approximately, by replacing in relation (29) the pressure angle a_1, with the normal pressure angle a_0 the relation taking the form (30); where ε_{12} represents the contact ratio of the gearing, and it will be calculated with expression (31) for the external gearing, and with relation (32) for the internal gearing.

$$\eta_m = \cfrac{1}{1 + tg^2\alpha_0 + \cfrac{2\pi^2}{3 \cdot z_1^2} \cdot (\varepsilon_{12} - 1) \cdot (2 \cdot \varepsilon_{12} - 1) \pm \cfrac{2\pi \cdot tg\alpha_0}{z_1} \cdot (\varepsilon_{12} - 1)} \quad (30)$$

$$\varepsilon_{12}^{a.e.} = \frac{\sqrt{z_1^2 \cdot \sin^2\alpha_0 + 4 \cdot z_1 + 4} + \sqrt{z_2^2 \cdot \sin^2\alpha_0 + 4 \cdot z_2 + 4} - (z_1 + z_2) \cdot \sin\alpha_0}{2 \cdot \pi \cdot \cos\alpha_0} \quad (31)$$

$$\varepsilon_{12}^{a.i.} = \frac{\sqrt{z_e^2 \cdot \sin^2\alpha_0 + 4 \cdot z_e + 4} - \sqrt{z_i^2 \cdot \sin^2\alpha_0 - 4 \cdot z_i + 4} + (z_i - z_e) \cdot \sin\alpha_0}{2 \cdot \pi \cdot \cos\alpha_0} \quad (32)$$

The made calculations have been centralized in the table 5.

Table 5

z1	α_0 [grad]	z2	ε_{12}^{ae}	η_{12}^{ae}	η_{21}^{ae}	ε_{12}^{ai}	η_{12}^{ai}	η_{21}^{ai}
			Determining the efficiency of the geared transmissions					
42	20	126	1.799	0.8447	0.8711	1.920	0.8386	0.8953
46	19	138	1.875	0.8567	0.8825	2.004	0.8509	0.9059
52	18	156	1.964	0.8693	0.8936	2.099	0.8640	0.9156
58	17	174	2.062	0.8809	0.9042	2.205	0.8760	0.9250
65	16	195	2.173	0.8921	0.9142	2.326	0.8876	0.9338
74	15	222	2.301	0.9033	0.9239	2.465	0.8992	0.9421
85	14	255	2.449	0.9140	0.9331	2.624	0.9104	0.9497
98	13	294	2.620	0.9242	0.9417	2.810	0.9209	0.9569
115	12	345	2.822	0.9340	0.9499	3.027	0.9312	0.9634
137	11	411	3.062	0.9435	0.9575	3.286	0.9410	0.9694
165	10	495	3.351	0.9522	0.9645	3.599	0.9501	0.9749
204	9	510	3.687	0.9607	0.9701	4.020	0.9586	0.9806
257	8	514	4.097	0.9684	0.9750	4.577	0.9662	0.9858
336	7	672	4.666	0.9753	0.9806	5.214	0.9736	0.9892
457	6	914	5.427	0.9818	0.9856	6.067	0.9802	0.9922
657	5	1314	6.495	0.9869	0.9898	7.264	0.9860	0.9946

Notations: ai=>inner gearings;
ae=>external gearing

8 Bended Teeth

Generally we use gearings with teeth inclined (with bended teeth). For gears with bended teeth, the calculations show a decrease in yield when the inclination angle increases. For angles with inclination which not exceed 25 degree the efficiency of gearing is good (see the table 6). When the inclination angle (β) exceeds 25 degrees the gearing will suffer a significant drop in yield (see the tables 7-8).

Table 6. *Bended teeth, β=25 [deg].*

Determining the efficiency when $\beta=25$ [deg]								
$z1$	α_0 [grad]	$z2$	ε_{12}^{ae}	η_{12}^{ae}	η_{21}^{ae}	ε_{12}^{ai}	η_{12}^{ai}	η_{21}^{ai}
42	20	126	1,708	0,829	0,851	1,791	0,826	0,871
46	19	138	1,776	0,843	0,864	1,865	0,839	0,883
52	18	156	1,859	0,856	0,876	1,949	0,853	0,895
58	17	174	1,946	0,869	0,889	2,043	0,866	0,906
65	16	195	2,058	0,882	0,900	2,151	0,879	0,917
74	15	222	2,165	0,894	0,911	2,275	0,892	0,927
85	14	255	2,299	0,906	0,922	2,418	0,904	0,936
98	13	294	2,456	0,917	0,932	2,584	0,915	0,945
115	12	345	2,641	0,928	0,941	2,780	0,926	0,953
137	11	411	2,863	0,938	0,950	3,013	0,937	0,961
165	10	495	3,129	0,948	0,958	3,295	0,947	0,968
204	9	510	3,443	0,957	0,965	3,665	0,956	0,974
257	8	514	3,829	0,965	0,971	4,146	0,964	0,981
336	7	672	4,357	0,973	0,977	4,719	0,972	0,985
457	6	914	5,064	0,980	0,983	5,486	0,979	0,989
657	5	1314	6,056	0,985	0,988	6,563	0,985	0,992

Table 7. *Bended teeth, β=35 [deg].*

			Determining the efficiency when $\beta=35$ [deg]					
z1	α_0 [grad]	z2	ε_{12}^{ae}	η_{12}^{ae}	η_{21}^{ae}	ε_{12}^{ai}	η_{12}^{ai}	η_{21}^{ai}
42	20	126	1,620	0,809	0,827	1,677	0,807	0,843
46	19	138	1,681	0,825	0,841	1,741	0,822	0,858
52	18	156	1,755	0,840	0,856	1,815	0,838	0,871
58	17	174	1,832	0,854	0,870	1,898	0,852	0,885
65	16	195	1,948	0,868	0,883	1,993	0,867	0,897
74	15	222	2,030	0,882	0,896	2,103	0,881	0,909
85	14	255	2,150	0,895	0,909	2,230	0,894	0,921
98	13	294	2,293	0,908	0,920	2,379	0,907	0,932
115	12	345	2,461	0,920	0,931	2,554	0,919	0,942
137	11	411	2,663	0,932	0,942	2,764	0,931	0,951
165	10	495	2,906	0,942	0,951	3,017	0,942	0,959
204	9	510	3,196	0,952	0,959	3,345	0,952	0,968
257	8	514	3,556	0,962	0,967	3,766	0,961	0,975
336	7	672	4,041	0,970	0,974	4,281	0,969	0,981
457	6	914	4,692	0,978	0,981	4,971	0,977	0,986
657	5	1314	5,607	0,984	0,986	5,942	0,984	0,990

Table 8. *Bended teeth, β=45 [deg].*

			Determining the efficiency when $\beta=45\ [deg]$					
z1	α_0 [grad]	z2	ε_{12}^{ae}	η_{12}^{ae}	η_{21}^{ae}	ε_{12}^{ai}	η_{12}^{ai}	η_{21}^{ai}
42	20	126	1,505	0,772	0,784	1,539	0,771	0,796
46	19	138	1,555	0,790	0,802	1,590	0,789	0,814
52	18	156	1,618	0,808	0,820	1,650	0,807	0,831
58	17	174	1,680	0,825	0,837	1,718	0,824	0,848
65	16	195	1,810	0,841	0,853	1,796	0,841	0,864
74	15	222	1,848	0,858	0,869	1,888	0,858	0,879
85	14	255	1,949	0,874	0,884	1,994	0,874	0,894
98	13	294	2,070	0,889	0,899	2,119	0,889	0,908
115	12	345	2,215	0,904	0,913	2,268	0,903	0,921
137	11	411	2,389	0,918	0,926	2,446	0,917	0,933
165	10	495	2,600	0,931	0,938	2,662	0,930	0,944
204	9	510	2,855	0,943	0,948	2,938	0,943	0,955
257	8	514	3,173	0,954	0,958	3,290	0,954	0,965
336	7	672	3,599	0,964	0,967	3,732	0,964	0,973
457	6	914	4,171	0,973	0,976	4,325	0,973	0,980
657	5	1314	4,976	0,981	0,983	5,161	0,981	0,986

New calculation relationships can be put in the forms (33-35).

$$\eta_m = \frac{z_1^2 \cdot \cos^2 \beta}{z_1^2 (tg^2\alpha_0 + \cos^2 \beta) + \frac{2}{3}\pi^2 \cos^4 \beta(\varepsilon-1)(2\varepsilon-1) \pm 2\pi tg\alpha_0 z_1 \cos^2 \beta(\varepsilon-1)} \quad (33)$$

$$\varepsilon^{a.e.} = \frac{1 + tg^2\beta}{2 \cdot \pi} \cdot$$
$$\cdot \left\{ \sqrt{[(z_1 + 2\cdot\cos\beta)\cdot tg\alpha_0]^2 + 4\cdot\cos^3\beta\cdot(z_1 + \cos\beta)} + \right. \quad (34)$$
$$+ \sqrt{[(z_2 + 2\cdot\cos\beta)\cdot tg\alpha_0]^2 + 4\cdot\cos^3\beta\cdot(z_2 + \cos\beta)} -$$
$$\left. - (z_1 + z_2)\cdot tg\alpha_0 \right\}$$

$$\varepsilon^{a.i.} = \frac{1 + tg^2\beta}{2 \cdot \pi} \cdot$$
$$\cdot \left\{ \sqrt{[(z_e + 2\cdot\cos\beta)\cdot tg\alpha_0]^2 + 4\cdot\cos^3\beta\cdot(z_e + \cos\beta)} - \right. \quad (35)$$
$$- \sqrt{[(z_i - 2\cdot\cos\beta)\cdot tg\alpha_0]^2 - 4\cdot\cos^3\beta\cdot(z_i - \cos\beta)} -$$
$$\left. - (z_e - z_i)\cdot tg\alpha_0 \right\}$$

The calculation relationships (33-35) are general. They have the advantage that can be used with great precision in determining the efficiency of any type of gearings.

To use them at the gearing without bended teeth is enough to assign them a beta value = zero. The results obtained in this case will be identical to the ones of the relations 30-32.

9 Conclusions

The best efficiency can be obtained with the internal gearing when the drive wheel 1 is the ring.

The minimum efficiency will be obtained when the drive wheel 1 of the internal gearing has external teeth.

For the external gearing, the best efficiency is obtained when the bigger wheel is the drive wheel.

When one decreases the normal angle α_0, the contact ratio increases and the efficiency increases as well.

References

1. Petrescu, R.V., Petrescu, F.I., Popescu, N.: Determining Gear Efficiency. Gear Solutions magazine, 19-28, March (2007);
2. Petrescu, F.I., *Theoretical and Applied Contributions About the Dynamic of Planar Mechanisms with Superior Joints* In online journal, "TesiOnline", 2009, Italy, http address:

   http://www.tesionline.com/intl/thesis.jsp?idt=262 87;
3. Petrescu, R.V., Petrescu, F.I., *CONTRIBUTIONS TO THE ANALYSIS AND SYNTHESIS OF MECHANISMS WITH BARS AND GEARING* - Book (in romanian), UniBook Publishing house, USA, January 2011, 218 pages.

Welcome

Annex

A brief history about the emergence and evolution of gearing mechanisms

A brief history about the emergence and evolution of gearing mechanisms

Top of the use of sprocket mechanisms must be sought in ancient Egypt with at least a thousand years before Christ. Here were used for the first time, transmissions wheeled "spurred" to irrigate crops and worm gears to the cotton processing.

With 230 years BC, in the city of Alexandria in Egypt, they have been used the wheel with more levers and gear rack.

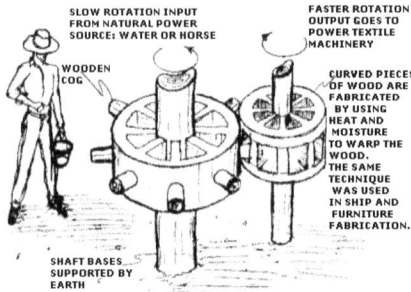

SLOW ROTATION INPUT FROM NATURAL POWER SOURCE: WATER OR HORSE

FASTER ROTATION OUTPUT GOES TO POWER TEXTILE MACHINERY

WOODEN COG

CURVED PIECES OF WOOD ARE FABRICATED BY USING HEAT AND MOISTURE TO WARP THE WOOD. THE SAME TECHNIQUE WAS USED IN SHIP AND FURNITURE FABRICATION.

SHAFT BASES SUPPORTED BY EARTH

Such gears have been constructed and used beginning from the earliest times, to the top for lifting the heavy anchors of vessels and for claim catapults used on the battlefields.

Then, they were introduced in cars with wind and water (as a reducing or multiplying at the pump from windmills or water).

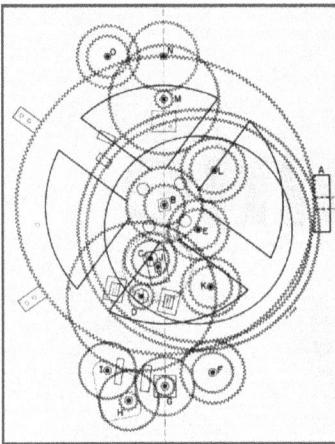

The Antikythera Mechanism is the name given to an astronomical calculating device, measuring about 32 by 16 by 10 cm, which was discovered in 1900 in a sunken ship just off the coast of Antikythera, an island between Crete and the Greek mainland. Several kinds of evidence point incontrovertibly to around 80 B.C. for the date of the shipwreck. The device, made of bronze gears fitted in a wooden case, was crushed in the wreck, and parts of the faces were lost, "the rest then being coated with a hard calcareous deposit at the same time as the metal corroded away to a thin core coated with hard metallic salts preserving much of the former shape of the bronze" during the almost 2000 years it lay submerged. (Antikythera 1).

It is hard to exaggerate the singularity of this device, or its importance in forcing a complete re-evaluation of what had been believed about technology in the ancient world. For this box contained some 32 gears, assembled into a mechanism that accurately reproduced the motion of the sun and the moon against the background of fixed stars, with a differential giving their relative position and hence the phases of the moon.

60

Modern adventure began with the gear wheel spurred of Leonardo da Vinci, in the fifteenth century. He founded the new kinematics and dynamics stating inter alia the principle of superposition of independent movements.

LEO's spurred wheel presented in detail. This CAR driven by a crank, is constructed simply, having a transmission consisting of a grooved wheel spurs and an axle. Spurs work on grooves, rotating axle. For the movement to be able to transmit, the axle has a groove in to a side.

Benz had engine with transmissions sprocket gearing and Gear chain (patented after 1882). On the right (up and down), you can see the drawings of a patent first gear transmission (first gearing patent) and of gearing wheels with chain made in 1870 by the British **Starley & Hillman.**

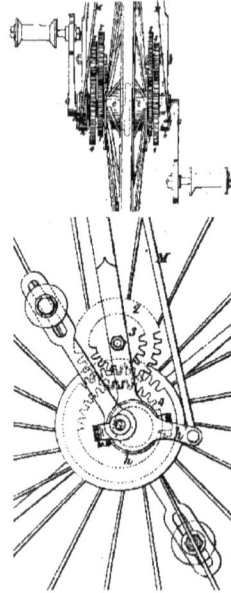

After 1912, in Cleveland (USA), begin to produce industrial specialized wheels and gears (cylindrical, worm, conical, with straight teeth, inclined or curved).

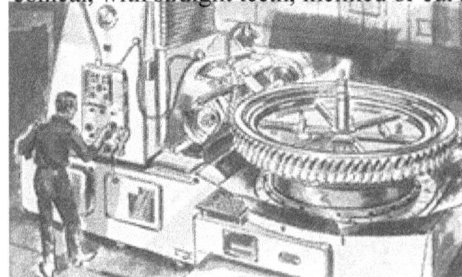

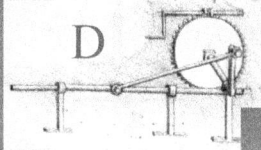

The old mechanisms with gearing (and bars) that were preserved:
A-ratchet;
B-worm screw mechanism and worm;
C-pendulum;
D-Leonardo da Vinci mechanism with worm, crank, rods and flies;
E-Planetary gears.

Gearing today.

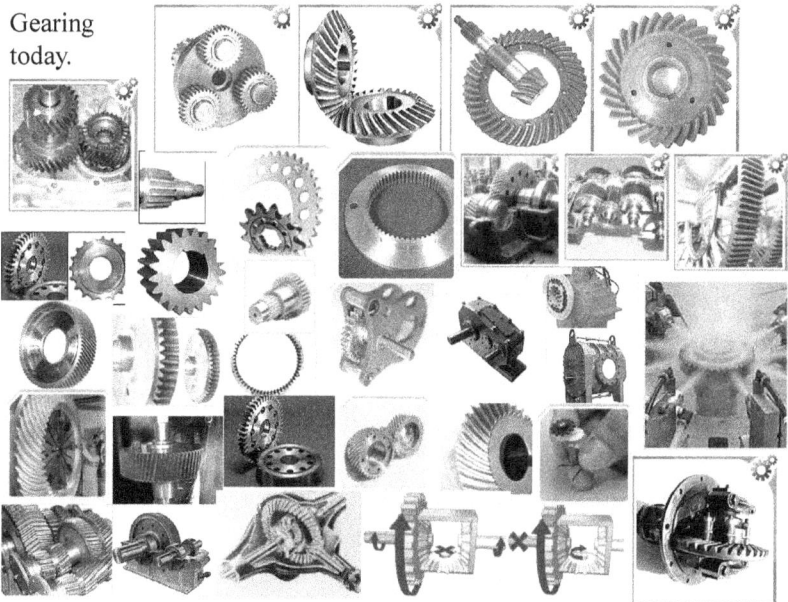

Gears and gearing for
heavy machinery.

Specialized gear reducers used in:

Aerospace Industry

Agricultural Industry

Auto Industry

Cement Industry

Naval industry

Mining Industry

Petrochemical Industry

Steel Industry

Sugar Industry

Materials Recycling Industry

Energy Industry

Paper Industry

Transmissions for railway and subway

Some areas of use gearing

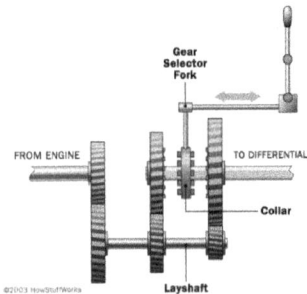 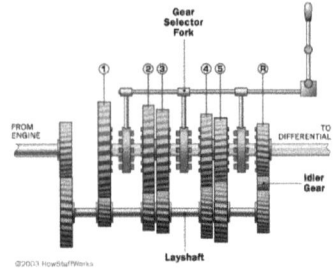

Classic Gearboxes (Manual)

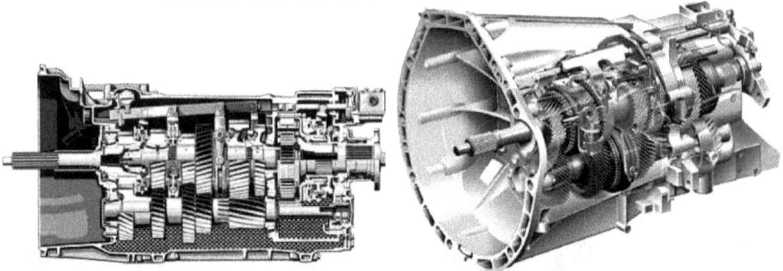

The first automatic transmission
(gears and planetary gears)

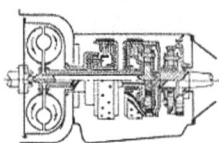 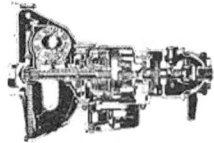 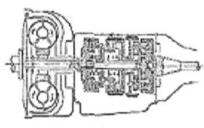

Fig. 1   Fig. 2   Fig. 3

Hydra-Matic transmission in Fig. 1, is an experienced model automatic transmission model Oldsmobile General Motors Corporation in 1940; The Dynaflow (Fig. 2), also created in 1948 by General Motors for Buick, was more effective;

Powerglide particular model, which was designed by General Motors in 1953, was a typical two-speed automatic transmission, which served as a standard model for other companies, so based on him, Ford's makes the Ford-O-Matic model (Fig. 3).

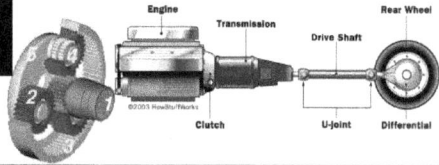

Automatic and modern gearboxes and CVTs

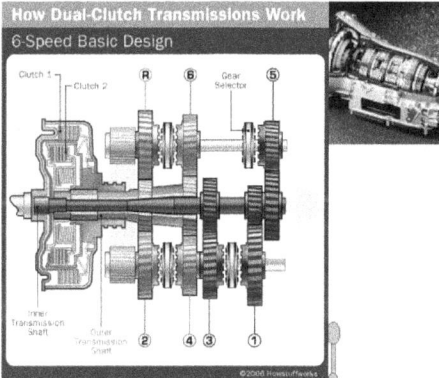

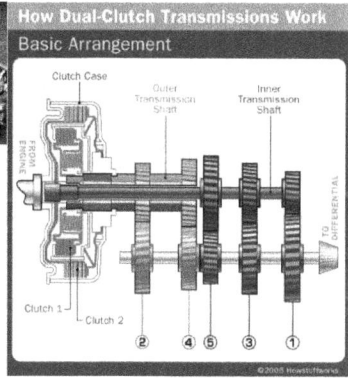

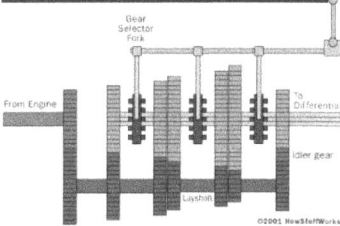

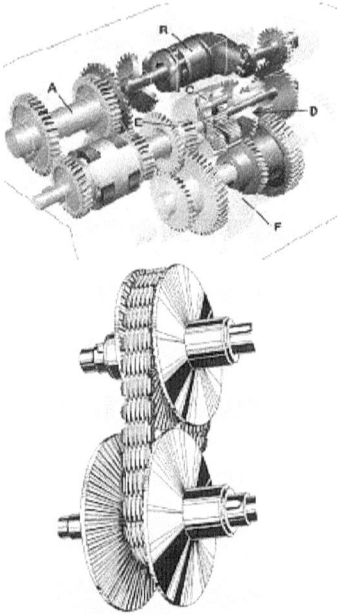

News
CVT classical or Variable Transmission (automatics) continuous (the classical model).

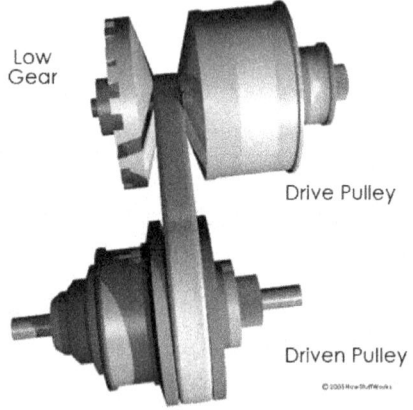

Low Gear

Drive Pulley

Driven Pulley

© 2005 HowStuffWorks

Gearbox exchanger swing mechanism. This mechanism, oscillating gear shifting is driven by a rotary cylinder, coupled to a lock mechanism (result is a high-speed gear shift). Oscillatory mechanism can be seen in next slide.

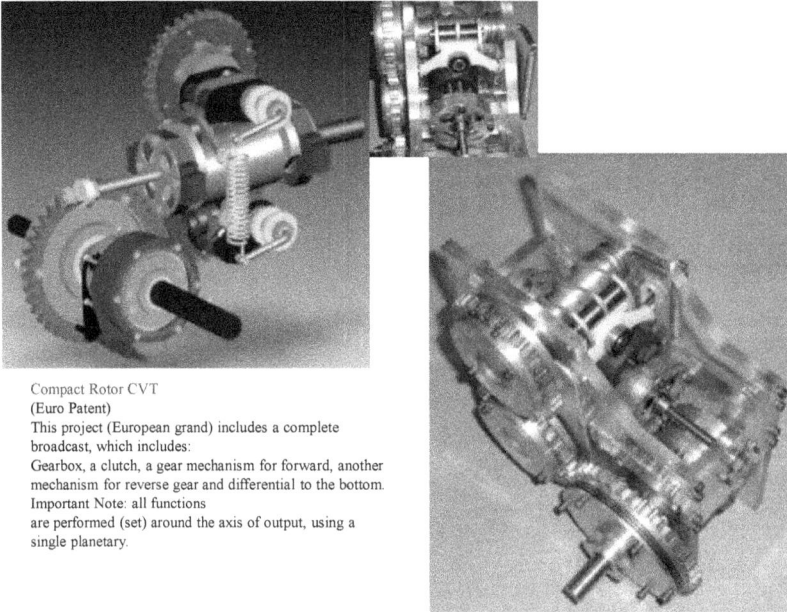

Compact Rotor CVT
(Euro Patent)
This project (European grand) includes a complete
broadcast, which includes:
Gearbox, a clutch, a gear mechanism for forward, another
mechanism for reverse gear and differential to the bottom.
Important Note: all functions
are performed (set) around the axis of output, using a
single planetary.

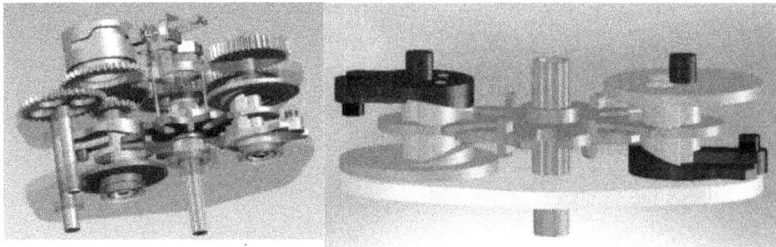

This patented mechanism of action for
exchangers speed sequential mechanism is based on a cross of Malta,
who works as a factor acting as the leader position for the gear change
soon.
This mechanism has the advantage of eliminating the command and the
hydraulic or servo, control, actuation and timing are all making with
Maltese cross mechanism changed.
The mechanism of action can be seen in next slide.

Actuator, the cross of Malta, amended.

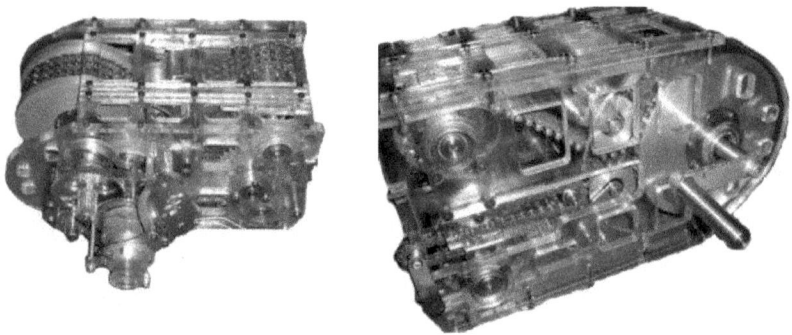

This compact model of CVT (continuously variable transmission), was presented at the "Frankfurt motor show, September 2007 and represents a fully functional model continuously variable transmission, compact, and autonomous. This model was exhibited at the exhibition organized and otherwise in the "SAE Commercial Vehicle Engineering Congress & Exhibition, October 30 - November 1, 2007 Rosemont (Chicago), Illinois, USA. Variable transmission mechanism is based on a drum, chain and bar.

Nautic
icvt

Perfect synchronization of
drive bars & chains

Variety of mechanisms
drum with gears, chain
and bar.

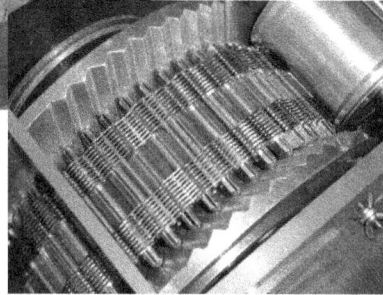